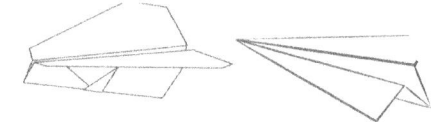

The Wright Brother's Ain't Got Nothing On Us, Paper edition!

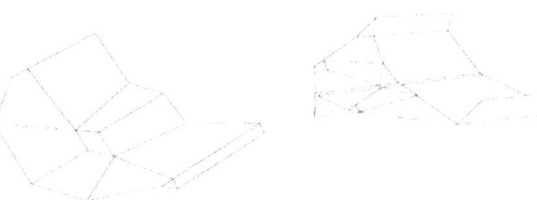

"How to make Paper Airplanes"

By: Mr.Sorgule's 1st period DDP class 2016-2017

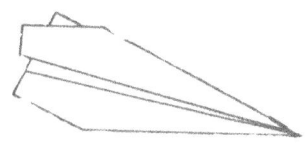
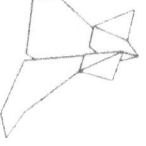

Your ISBN: 978-1-387-02463-6

The Headless Horseman

Result of step 1

Step-1

Result of step 2

Step-2

Step-3

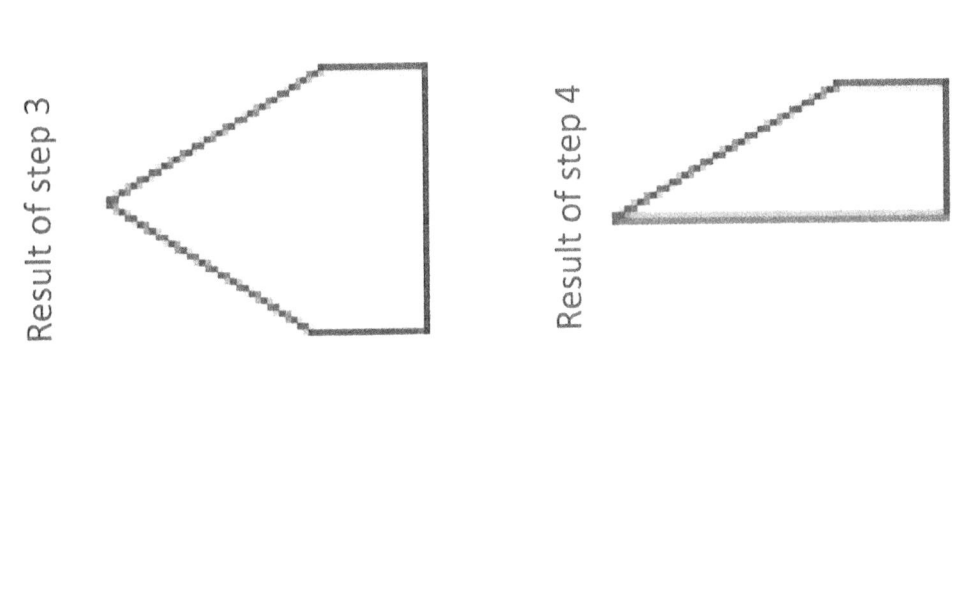

Result of step 3

Step-4

Result of step 4

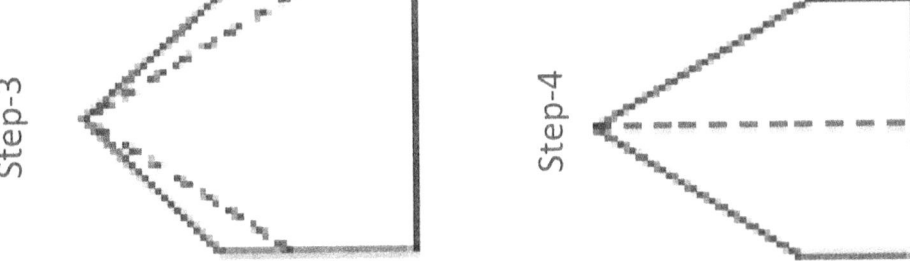

Result of step 5

Result of step 6

Step-5

Step-6

BIPLANE

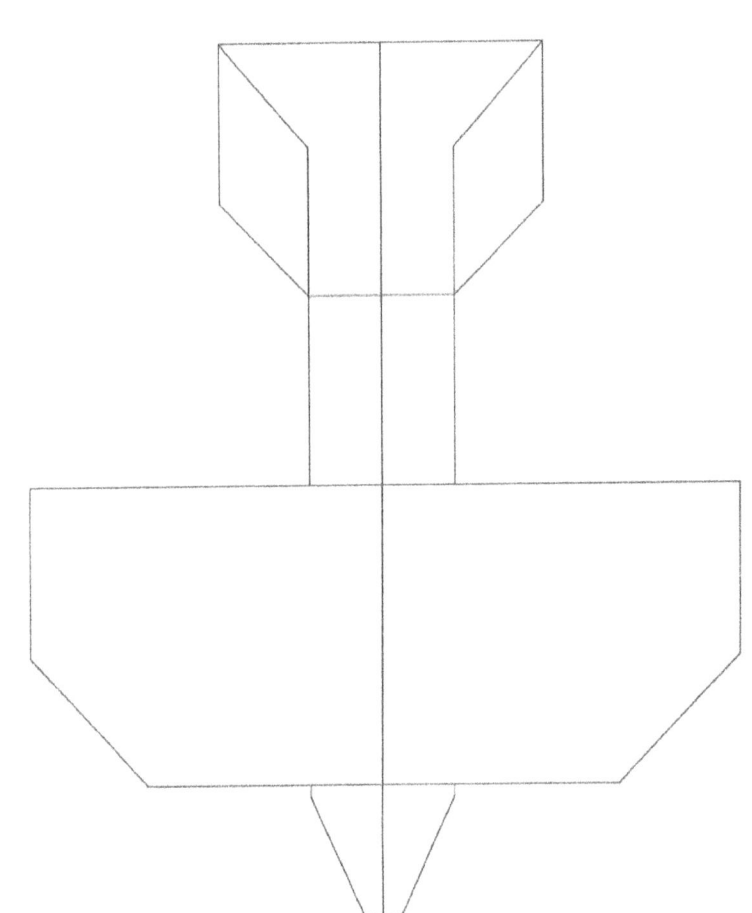

You will need 2 sheets of 8 ½ x 11 inch (A4-size) paper and a penny. You don't need scissors.

The Body

The first thing we will be working on is the body. The wings come after. The body and the wings **DO _NOT_** require (need) any cutting.

Step: 1

First hot dog fold the paper.
Valley fold the paper one
quarter of the way in.

1/4

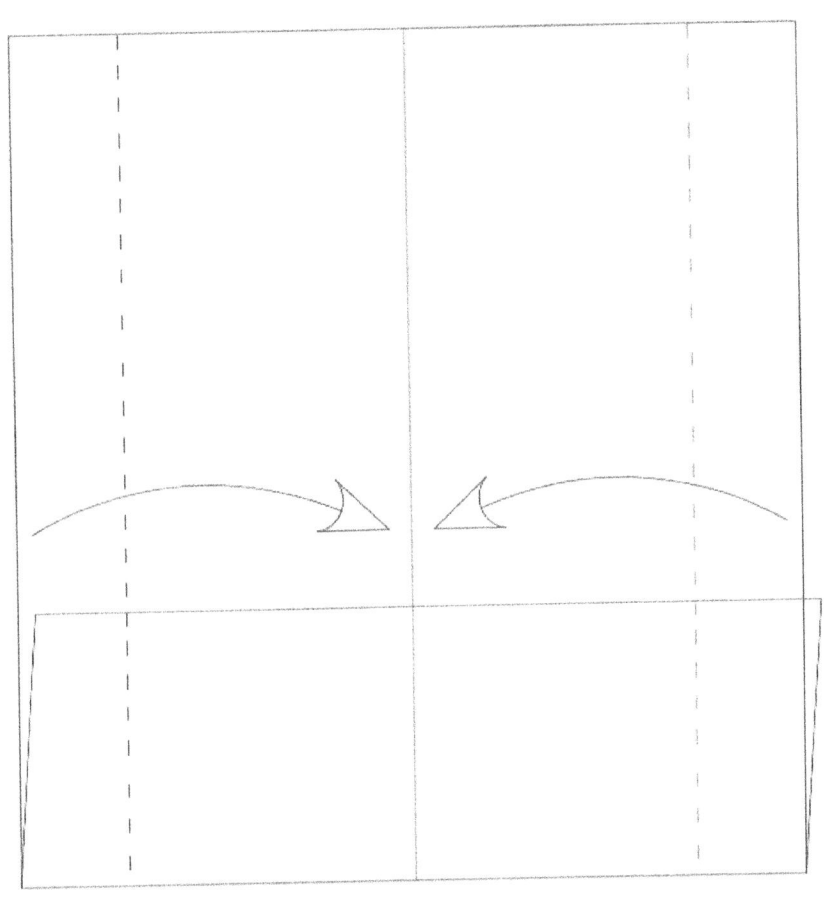

Step: 2

Fold on the dotted lines.
Then valley fold both
sides into the center.

Step: 3
Valley fold back out to the edge.

Keagen LaGrave

The Hammerhead

The First step is to fold the paper directly down the middle.

The second step fold the top two corners.

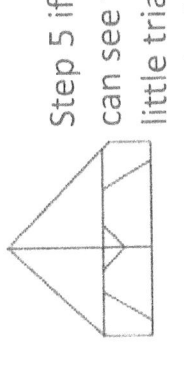

Step 3 now you will fold the paper down. You will fold it a little under the other one.

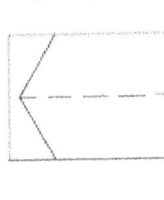

Step 4 now that its folded now fold the top two corners again.

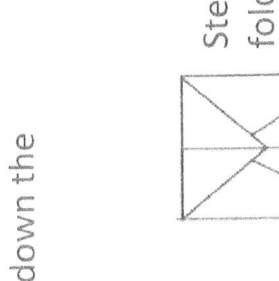

Step 5 if you can see the little triangle on the bottom and fold that up.

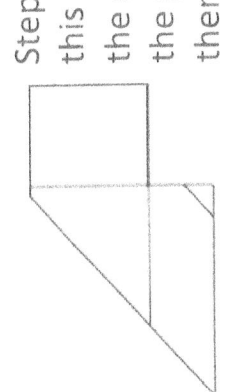

Step 6 For this step you will fold this side as show and do this to the other side. Once you do that the plan is able to fold in and then pull out.

The zump / twirler

John LeClair

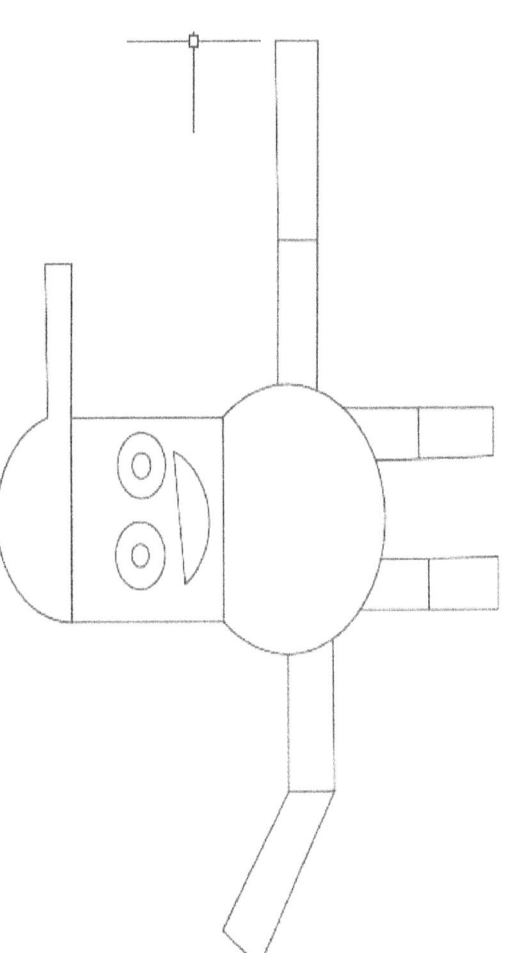

1. Then take one of the halves and fold it in half

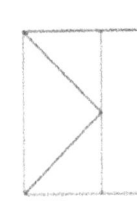

2. Then fold the corners inwards

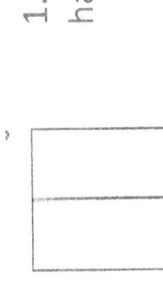

3. Fold down the corners made by the creases

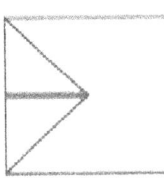

4. Then fold the pre folded corners up

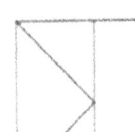

5. Fold the upper large triangle in half

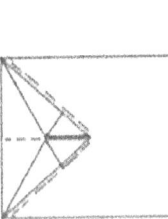

6. Fold down both wings, approximately where indicated its wings are going to be smaller than the body

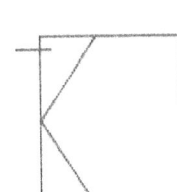

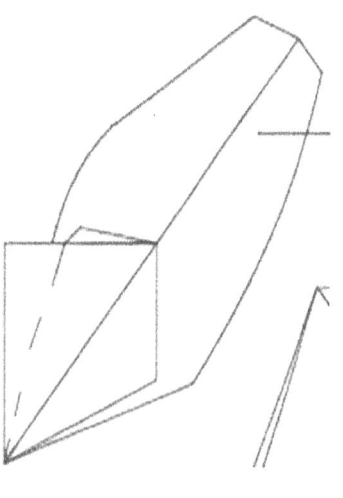

8. Now fold the center triangle in half, tuck it in, and fold the paper airplane flat again

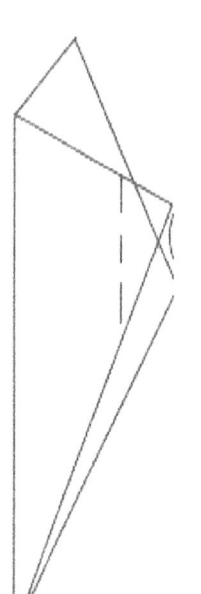

9. Crimp the wingtips as shown and lift the wings up so that they are angled slightly upward and then its done

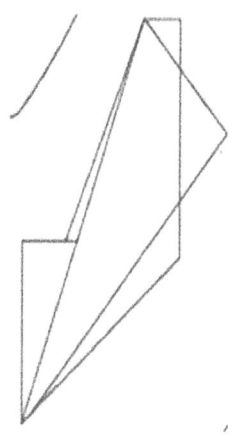

7. Carefully open the paper airplane, keeping the center triangle together and to one side, with the smaller flap held against the opposite side as in diagram 12

Basic stunt plane

Edward Nephew

Step5:fold wings 1" up from bottom of plane

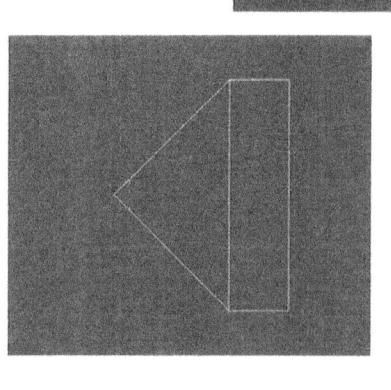

Step2: repeat step one with first fold done

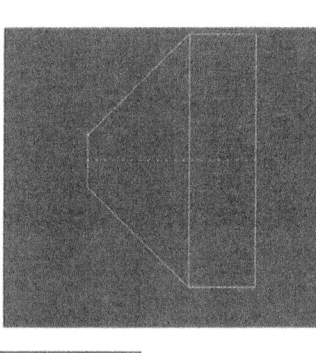

step4:fold down center

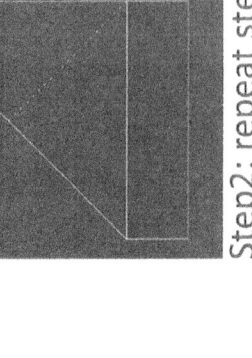

Step3:fold tip 1" from top point

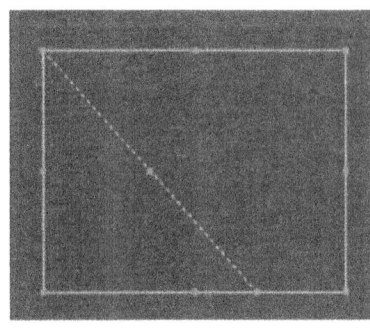

Step1: Fold a 8.5" by 11" piece of paper so that one corner lines up to the side edge

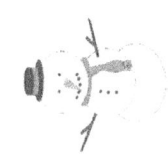

The Flyer

Fold on the dotted line.

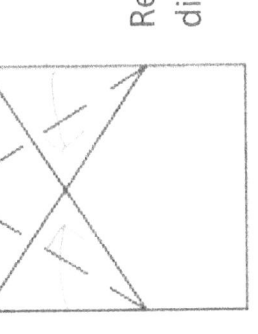

Repeat the first direction.

I bet you cant guess what you have to do now.. Fold on the dotted lines again.

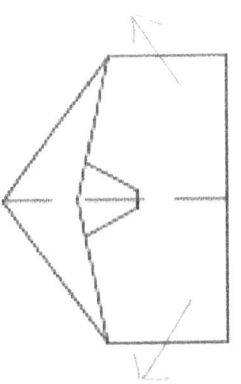

Still folding the dotted lines.. How fun?

Fold it hot dog style!

I hate to say it.. But fold on the dotted line again.

No more dotted lines!!.

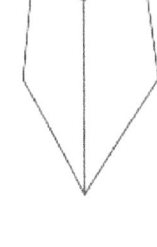

Tape.. And your hands..

Danielle Reynolds

The Speed plane

By: Tyler Gadway,

Fold in half for a crease to start on

Fold corners in crease for four times

Fold the sides for the wings

How to make an underside plane

starring the Assension man

Step 1: Get paper 8.5x11 inches

Step 2: Fold in half like shown

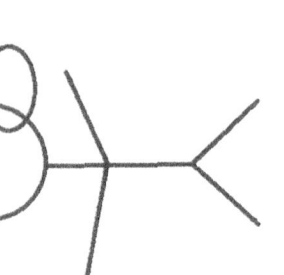

Step 3: Fold the top down

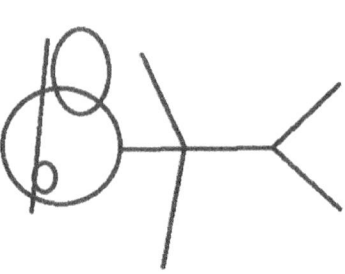

Step 4: Fold the top down again

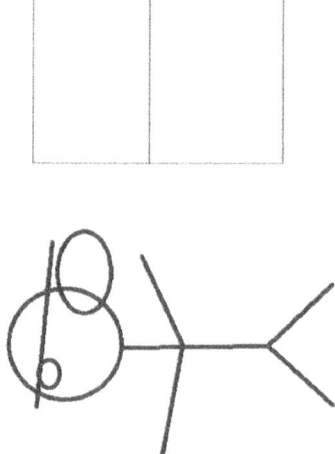

Step 5: Fold a corner in

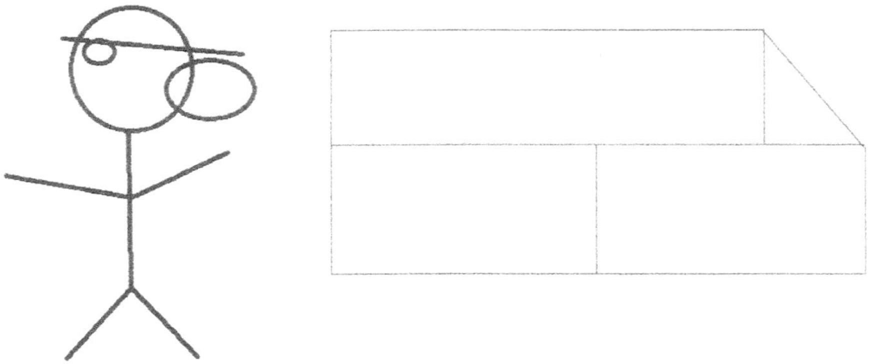

Step 6: Fold in the other corner

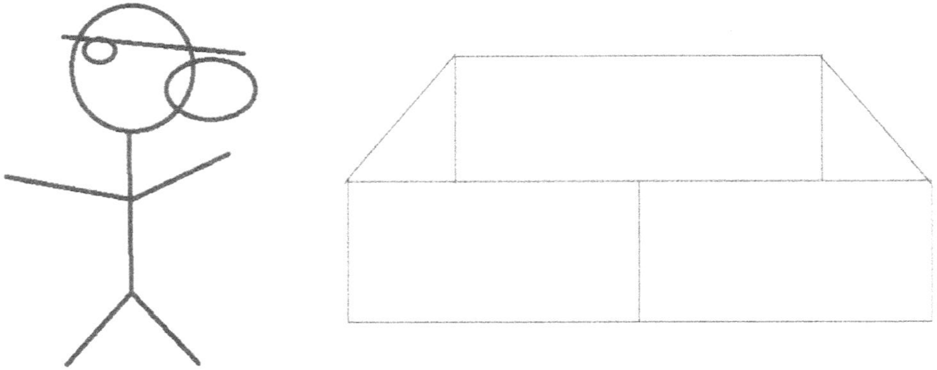

Step 7: Fold the corners in once again

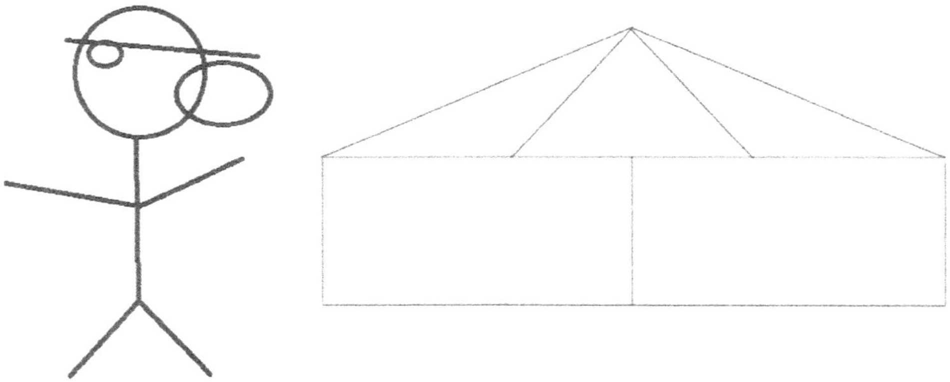

Step 8: Fold down the top

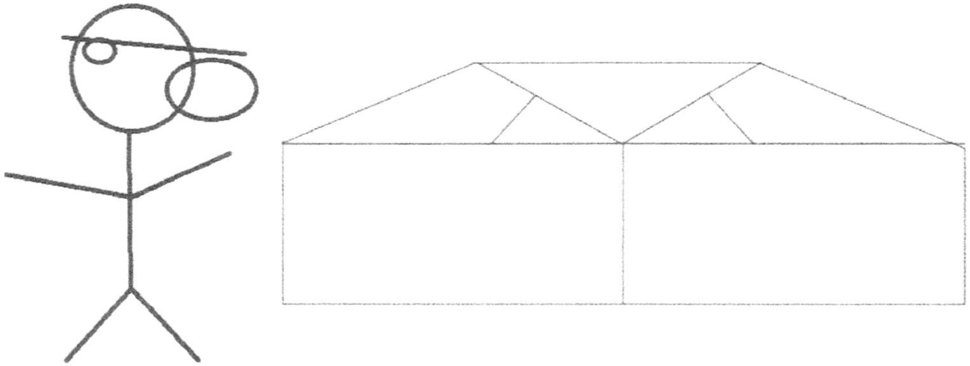

Step 9: Fold the sides into each other

Step 10: Fold the wings about 1 inch from the bottom

Now test it out in a room with easily breakable objects

The Hunting Flight

Step one

Fold The paper in half lengthwise, then unfold.

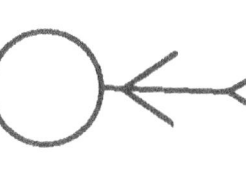

Step two

Fold the top down approximately two inches.

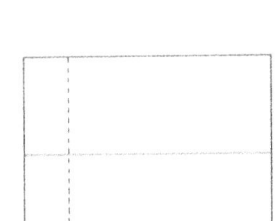

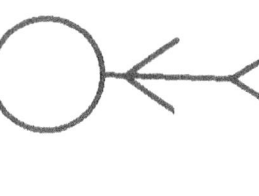

Step three

Fold the top down one inch.

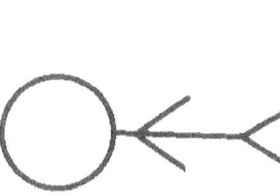

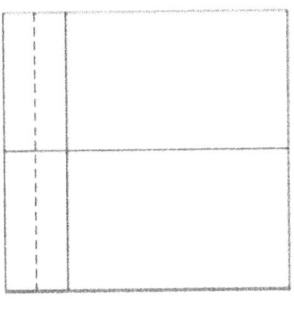

Nicholas Szczerbak

Step four

Fold the top down once more, but only half an inch this time.

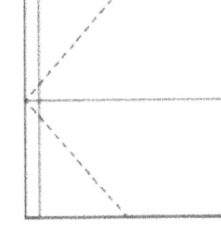

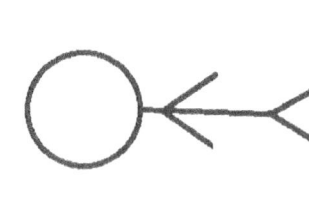

Step five.

Fold the top corners inward, toward the middle to give the top a triangle shape.

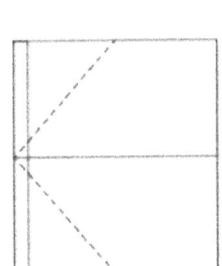

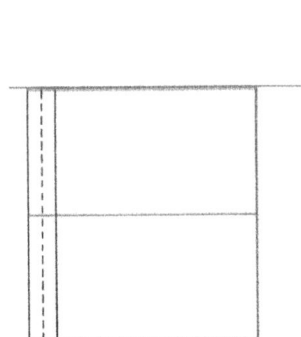

Step six

Fold the plane in half so the thick part of the paper is on the outside.

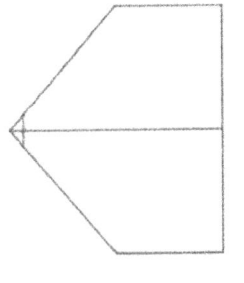

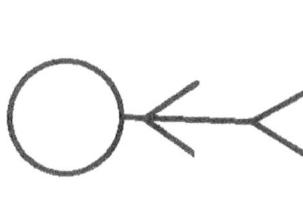

Step seven

Fold the wings down at a slight angle, then fold up the edges of the wings.

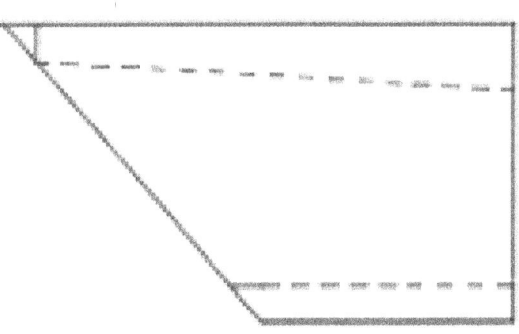

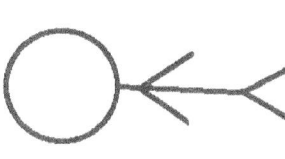

The finished plane.

Your plane should look similar to these front, top and side views.

This plane Can be thrown with any force, it flies extremely well with gentle to medium throws.

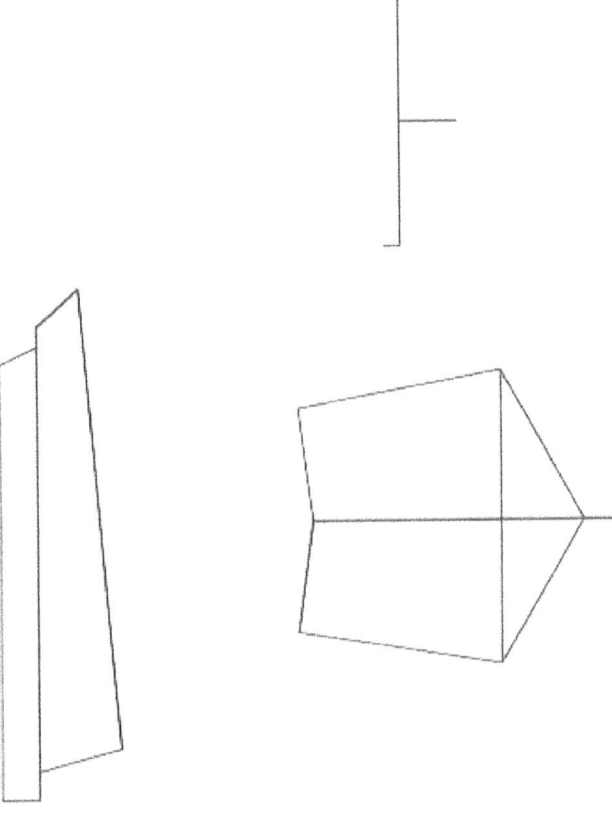

MATT'S COOL PAPER AIRPLANE!

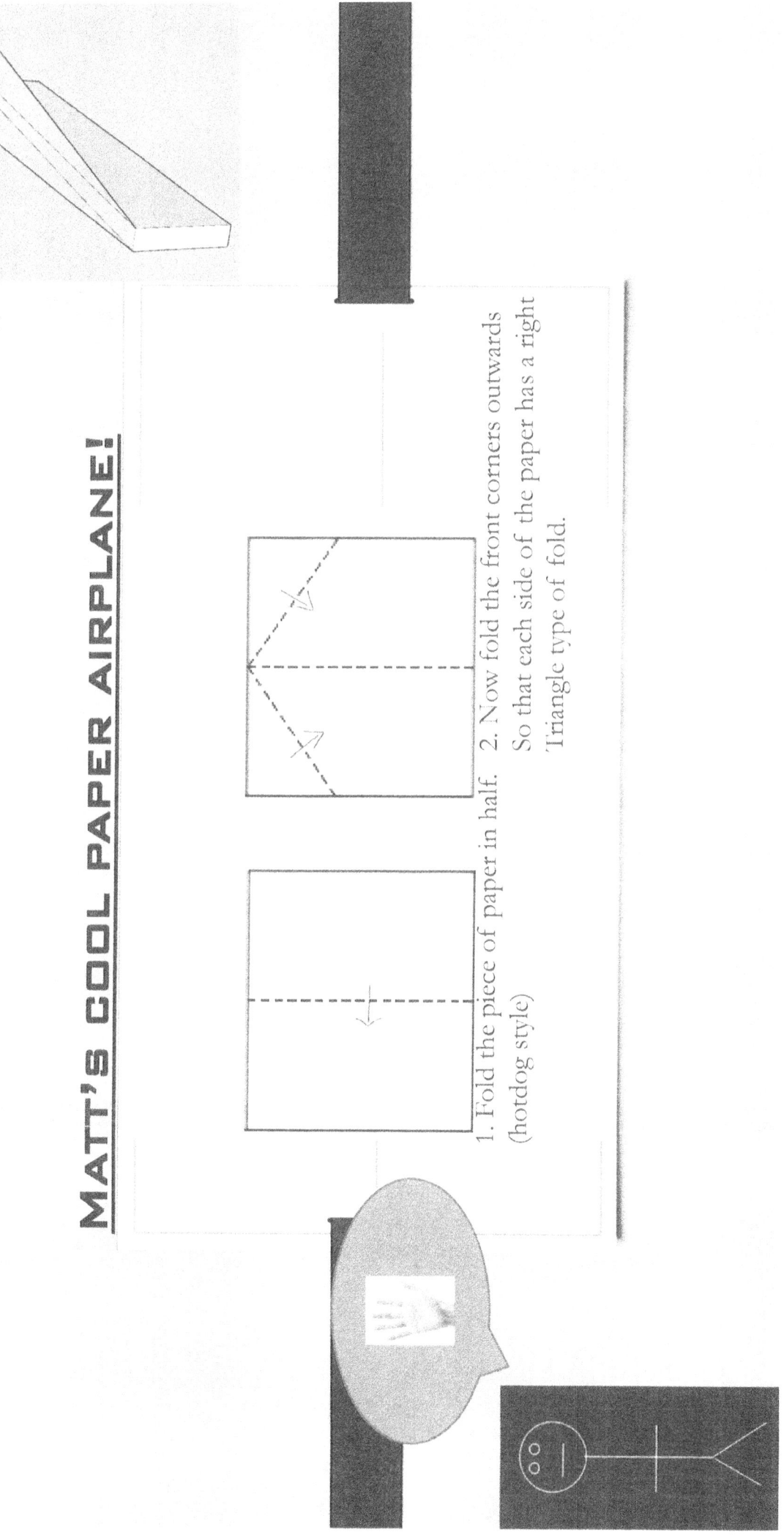

1. Fold the piece of paper in half. (hotdog style)

2. Now fold the front corners outwards So that each side of the paper has a right Triangle type of fold.

MATT'S COOL PAPER AIRPLANE!

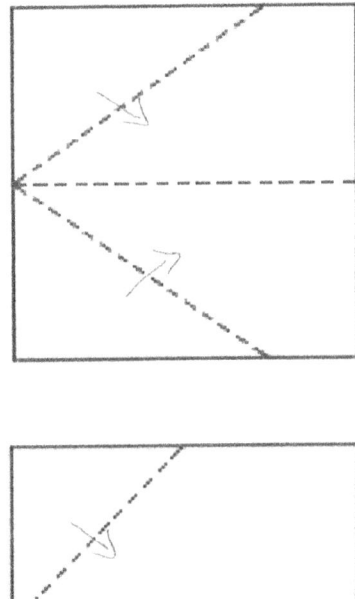

3. Repeat step 2

4. Repeat step 2

By: Matt Forrence

Erick Frechette

THE TANK

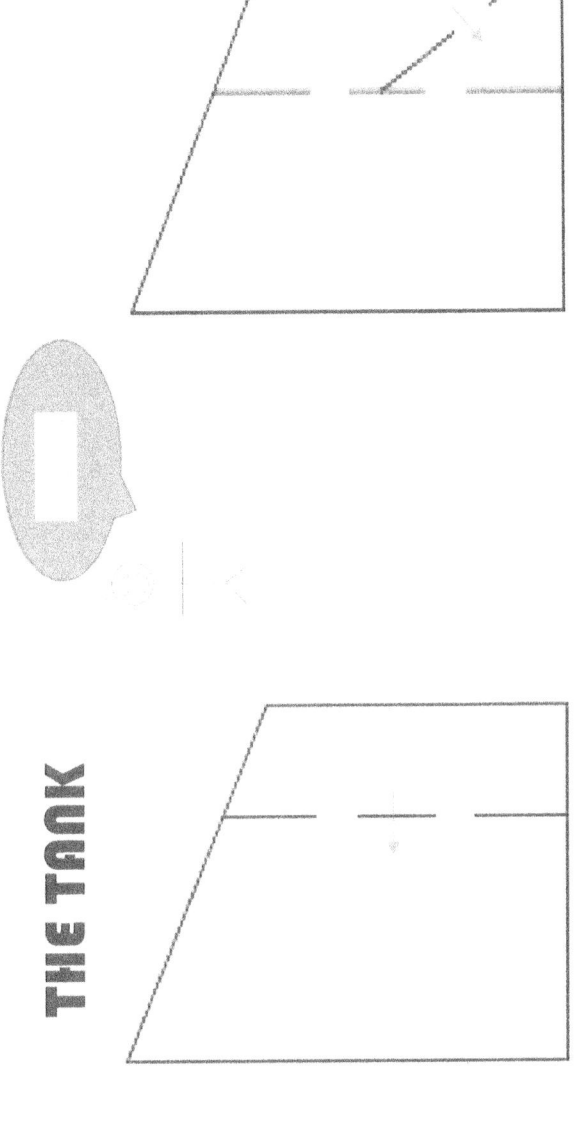

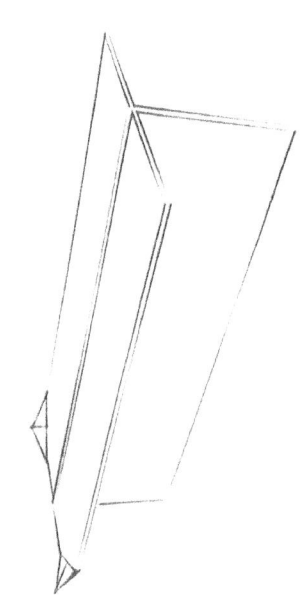

1. Fold the top pieces down.
2. Fold the paper in half long Ways.

1. Fold the wing of the Plane down the put it Sideways.

1. Fold the piece on the Wing upwards.

Regan Baker The mini paper Airplane.

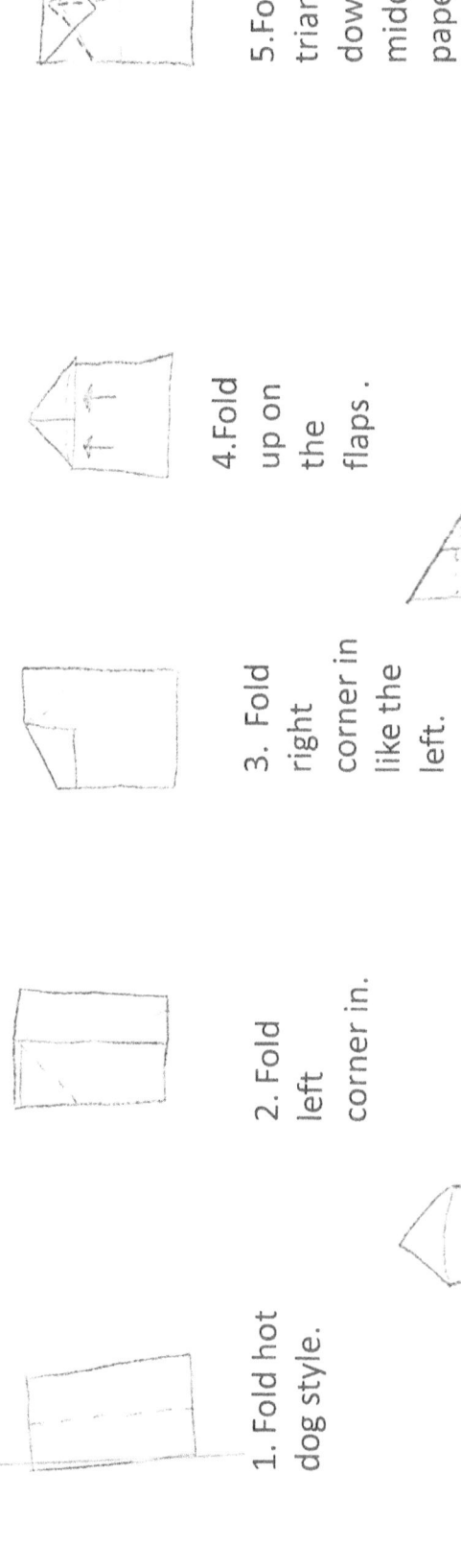

1. Fold hot dog style.

2. Fold left corner in.

3. Fold right corner in like the left.

4. Fold up on the flaps .

5. Fold triangle down to middle of paper.

6. Should have a triangle shape at top of paper.

7. Fold the corner out to make wings.

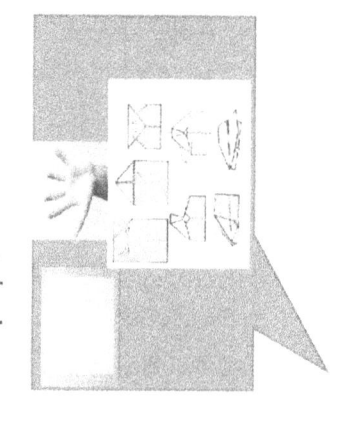
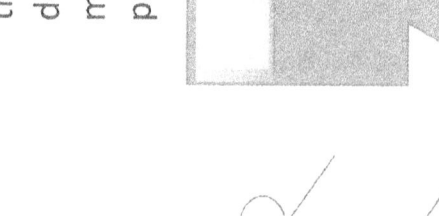
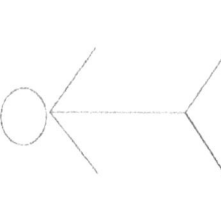
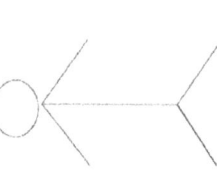
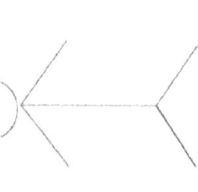

By:
Jacob
Breen

Thank you for flying
Paper delta!

2.

4.

1.

First take paper and fold in half.

Then fold the top corners of your paper down the your crease.

3.

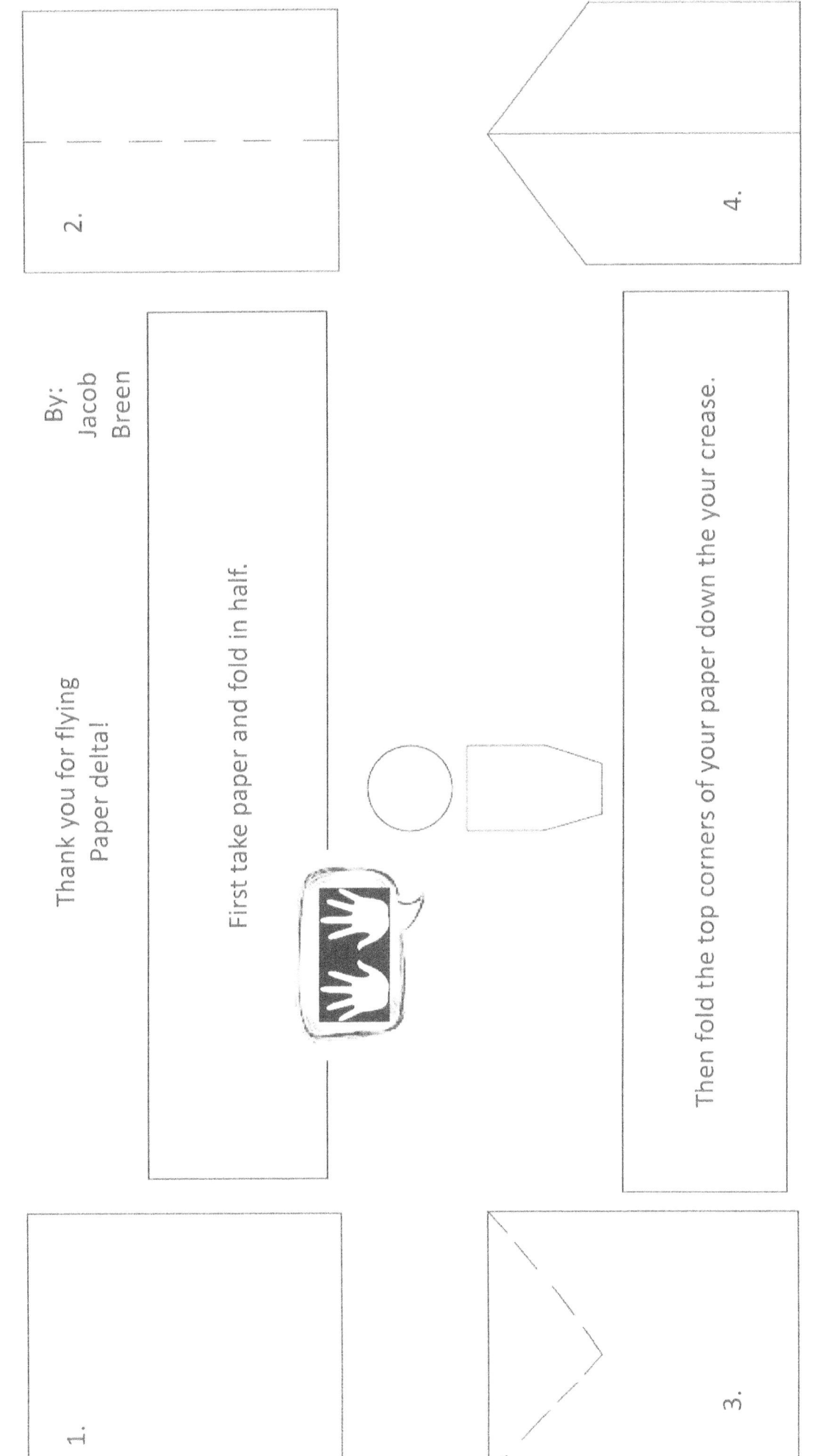

Then refold that paper back in half and take your top right corner and fold that to the left side.

6.

5.

Don't mess up!

The go back one step and bring your top farthest right and left corner and fold it to the middle.

8.

7.

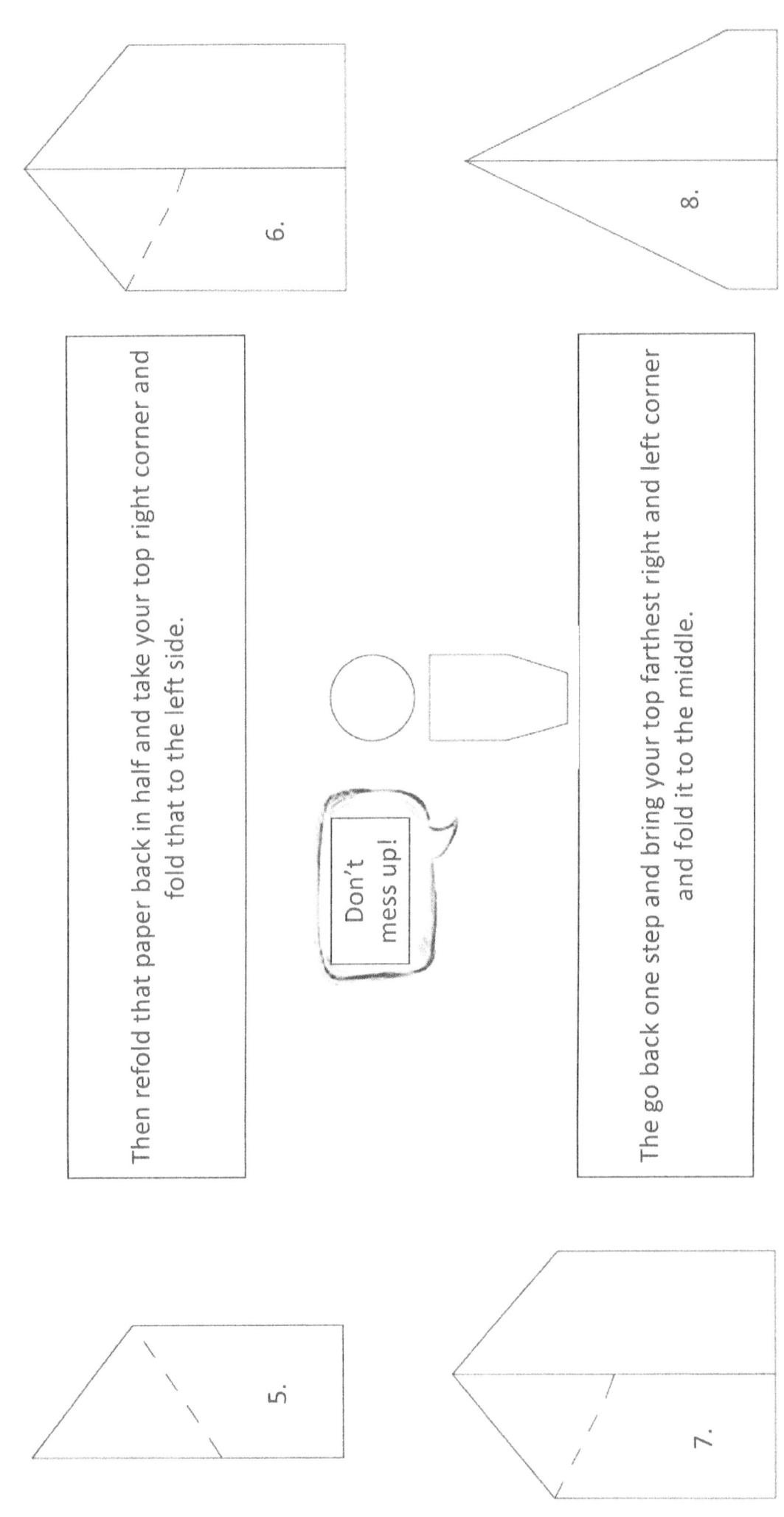

Then fold that in half and your finished.

Now throw and watch it fly.

9.

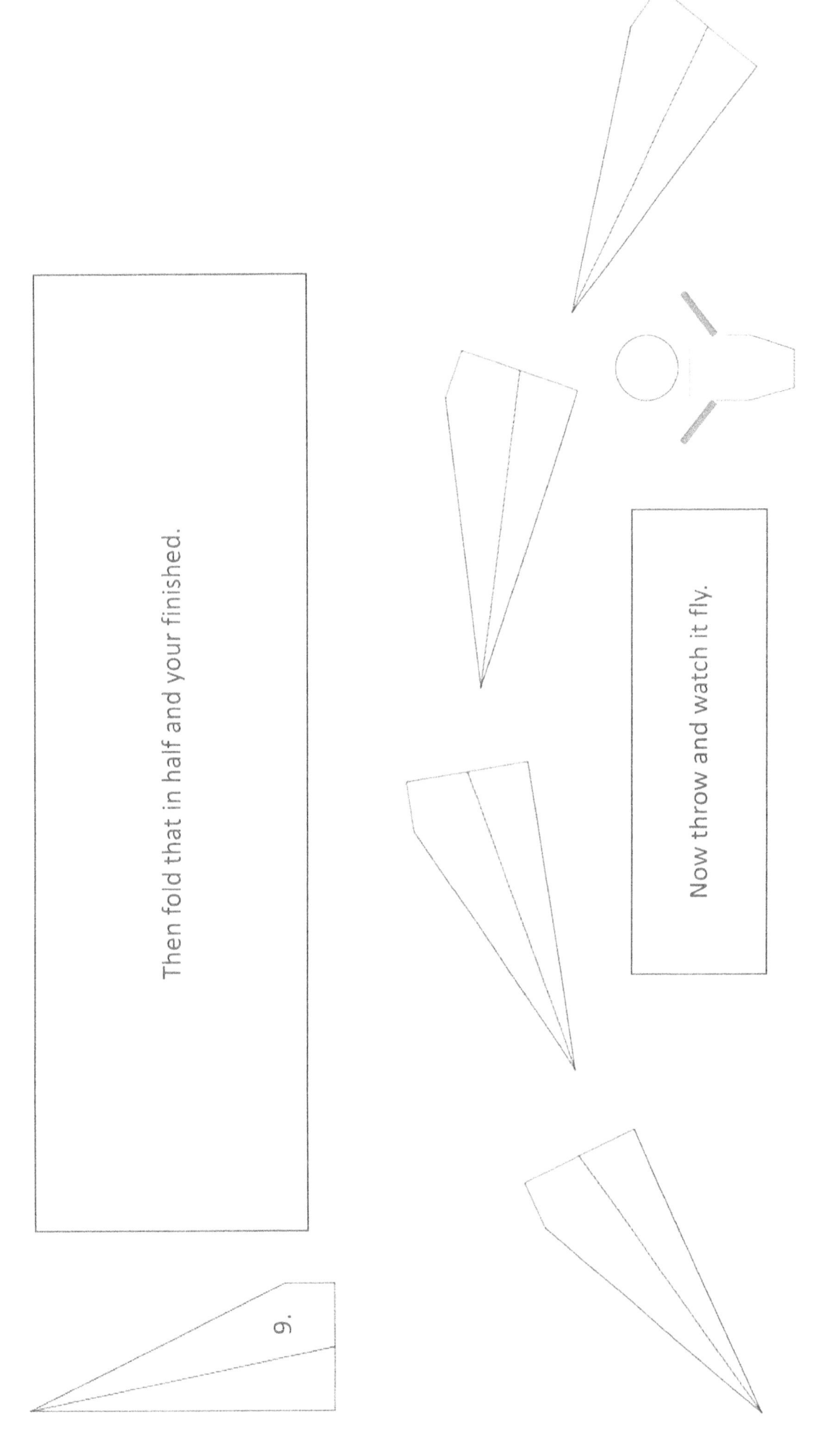

The Flyer

By Emily Beattie

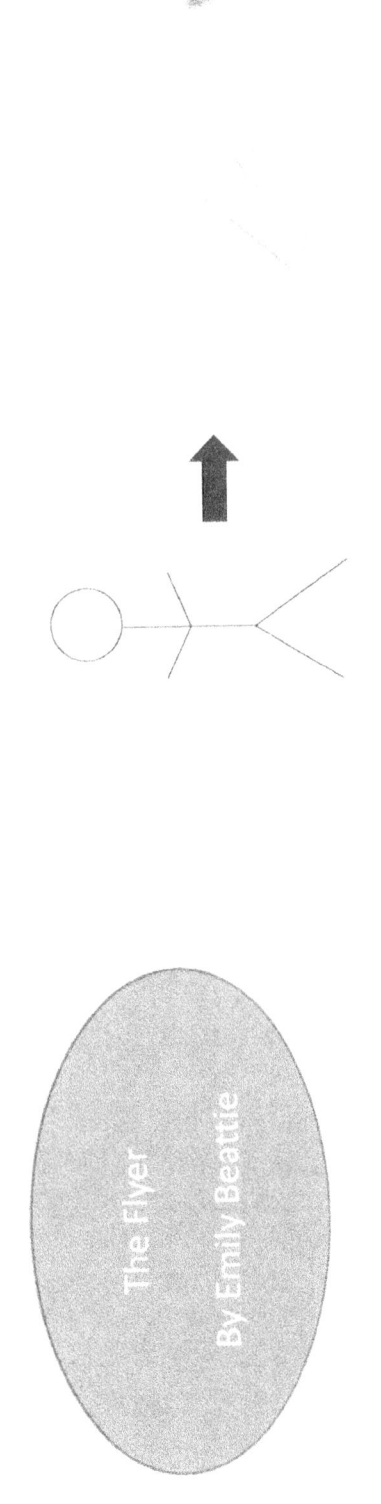

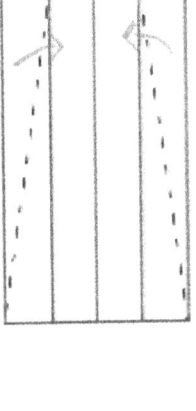

Step 1

Fold Paper in half "hot dog" style

Step 2

Fold both half of the half to the middle

Step 3

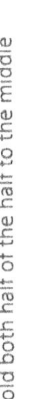

Fold top corners to the middle of the top of the page

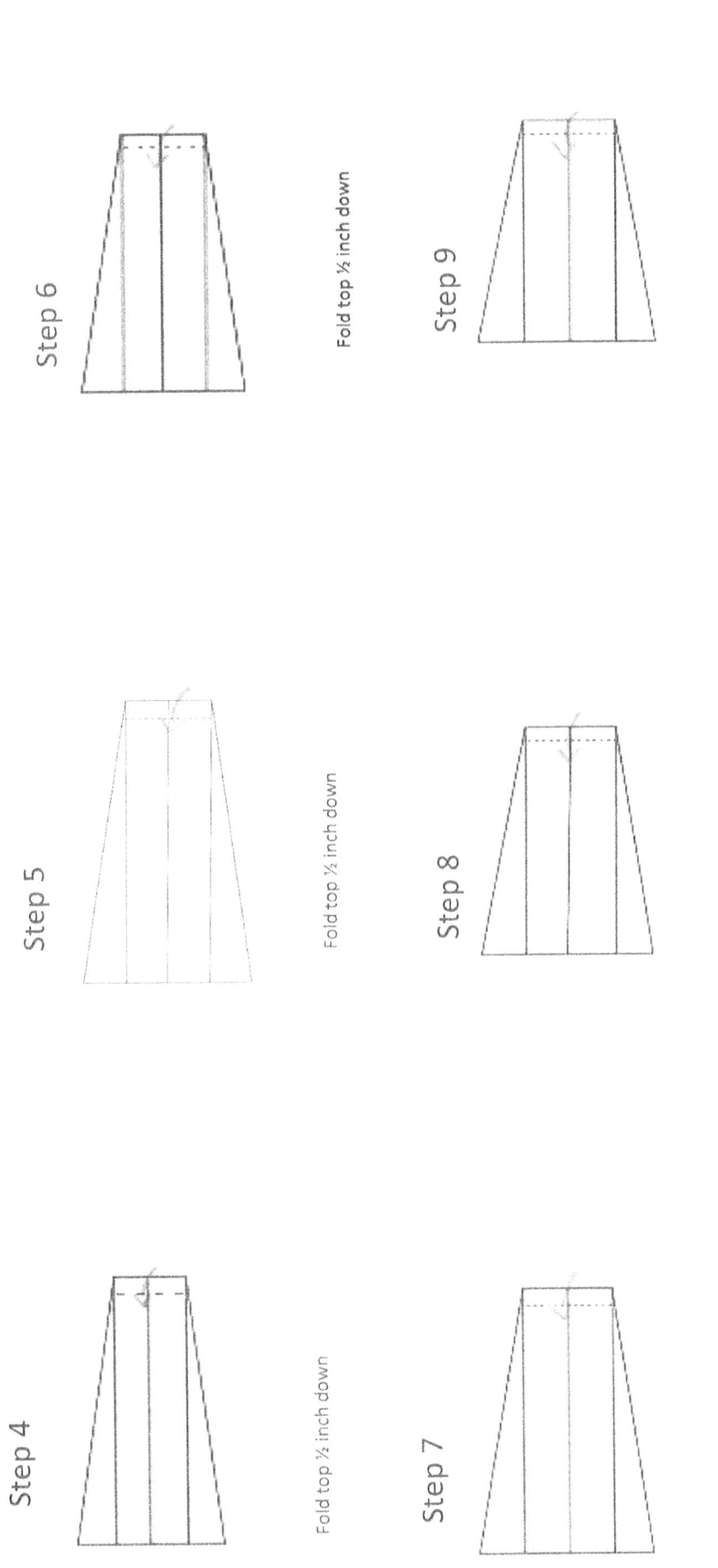

Step 4

Fold top ½ inch down

Step 5

Fold top ½ inch down

Step 6

Fold top ½ inch down

Step 7

Fold top ½ inch down

Step 8

Fold top ½ inch down

Step 9

Fold top ½ inch down

Step 10

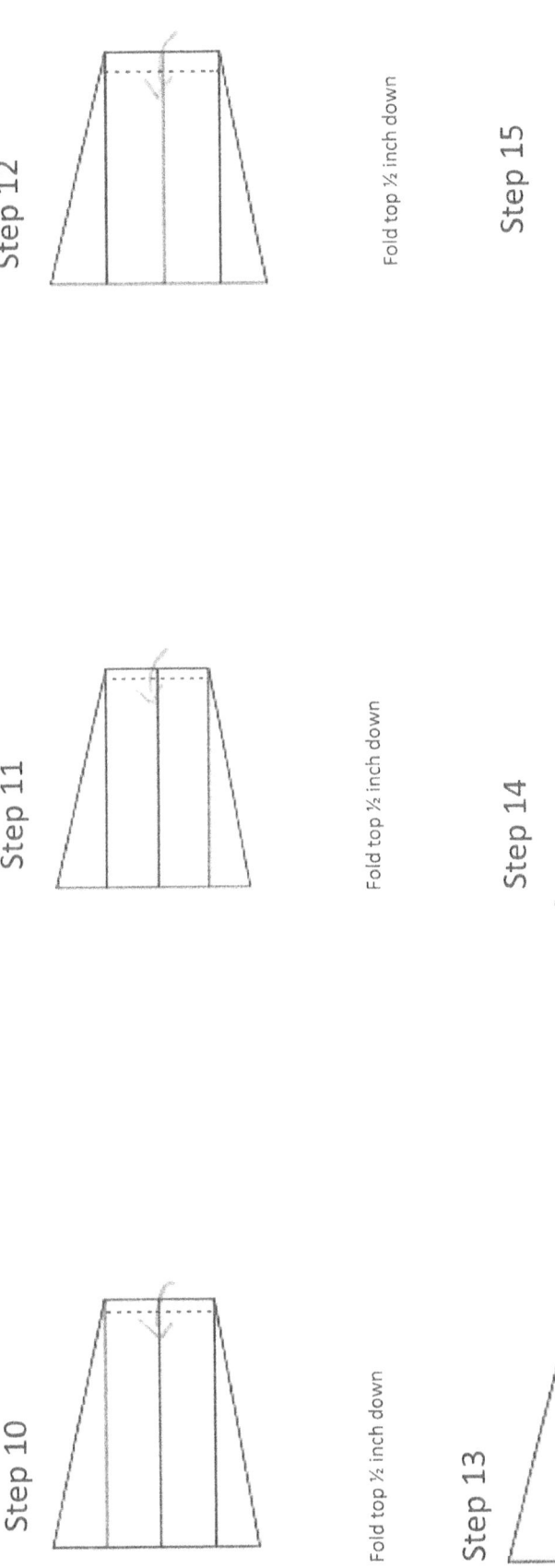

Fold top ½ inch down

Step 11

Fold top ½ inch down

Step 12

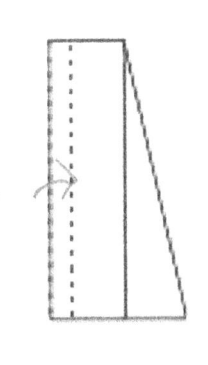

Fold top ½ inch down

Step 13

Fold paper back in half

Step 14

Fold edge of paper ½ inch in

Step 15

Flip paper over and do the same thing to the other side

Step 16

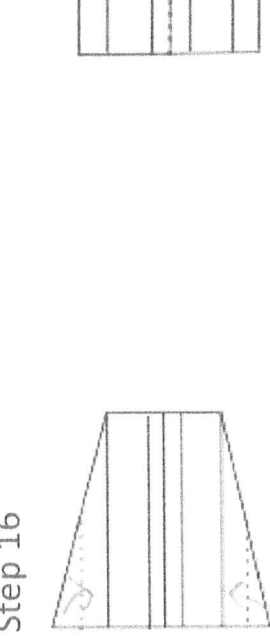

Fold corners in

Final

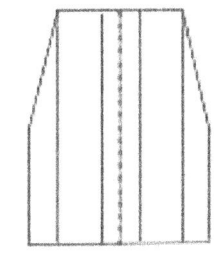

Final Project

Created by: Keith Schubert

THE MALCOM

1. Fold left or right then Fold back

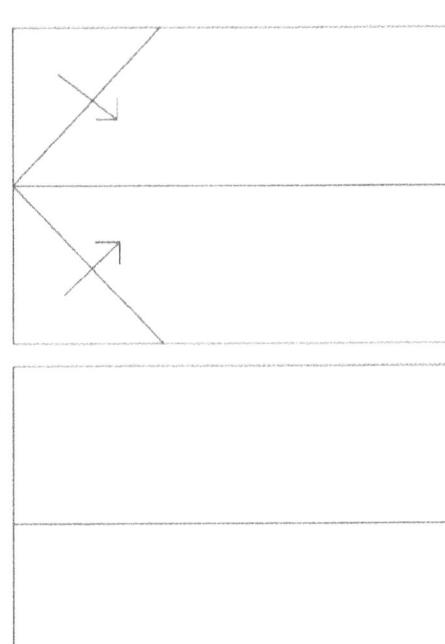

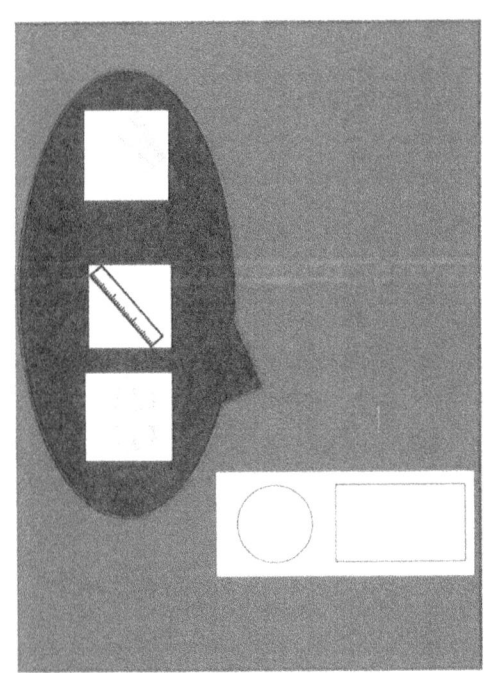

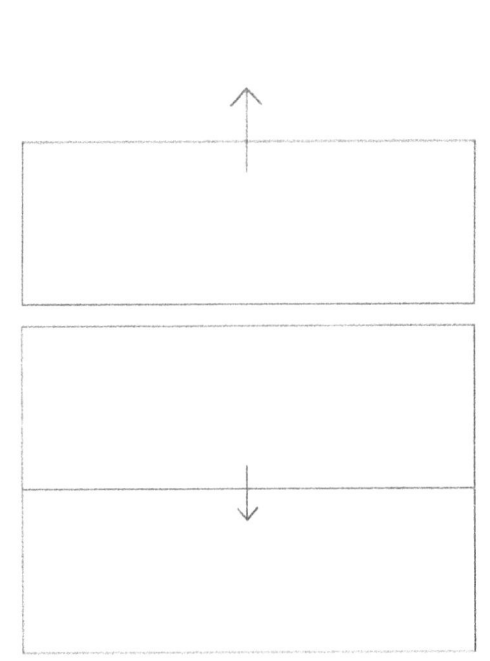

2. Fold the corners

3. Fold it sideways

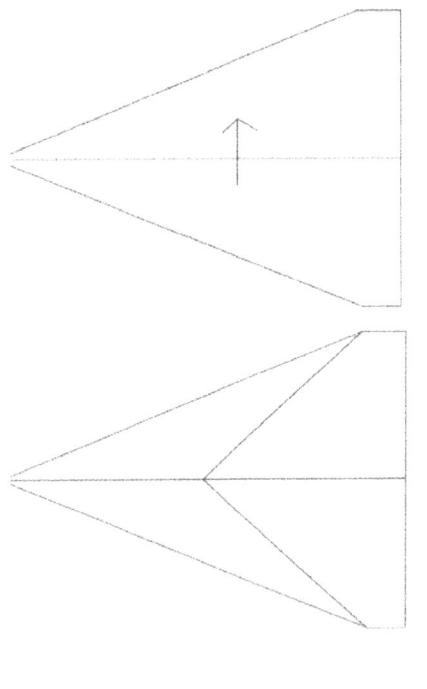

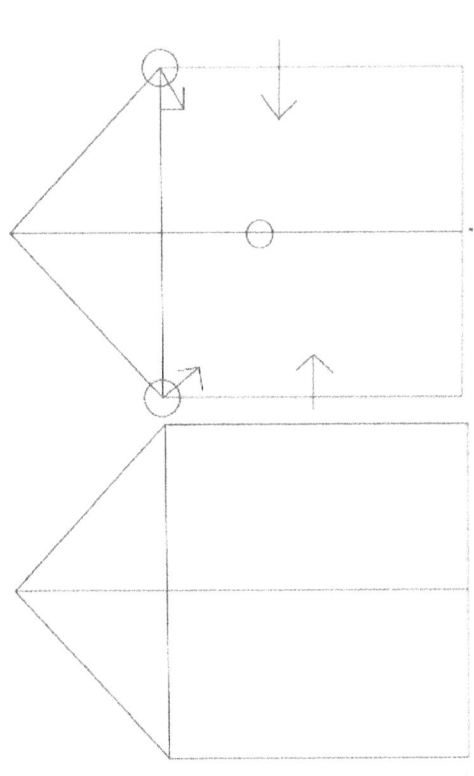

4. Flip over to other side

5. Fold down on both sides

6. Final Product

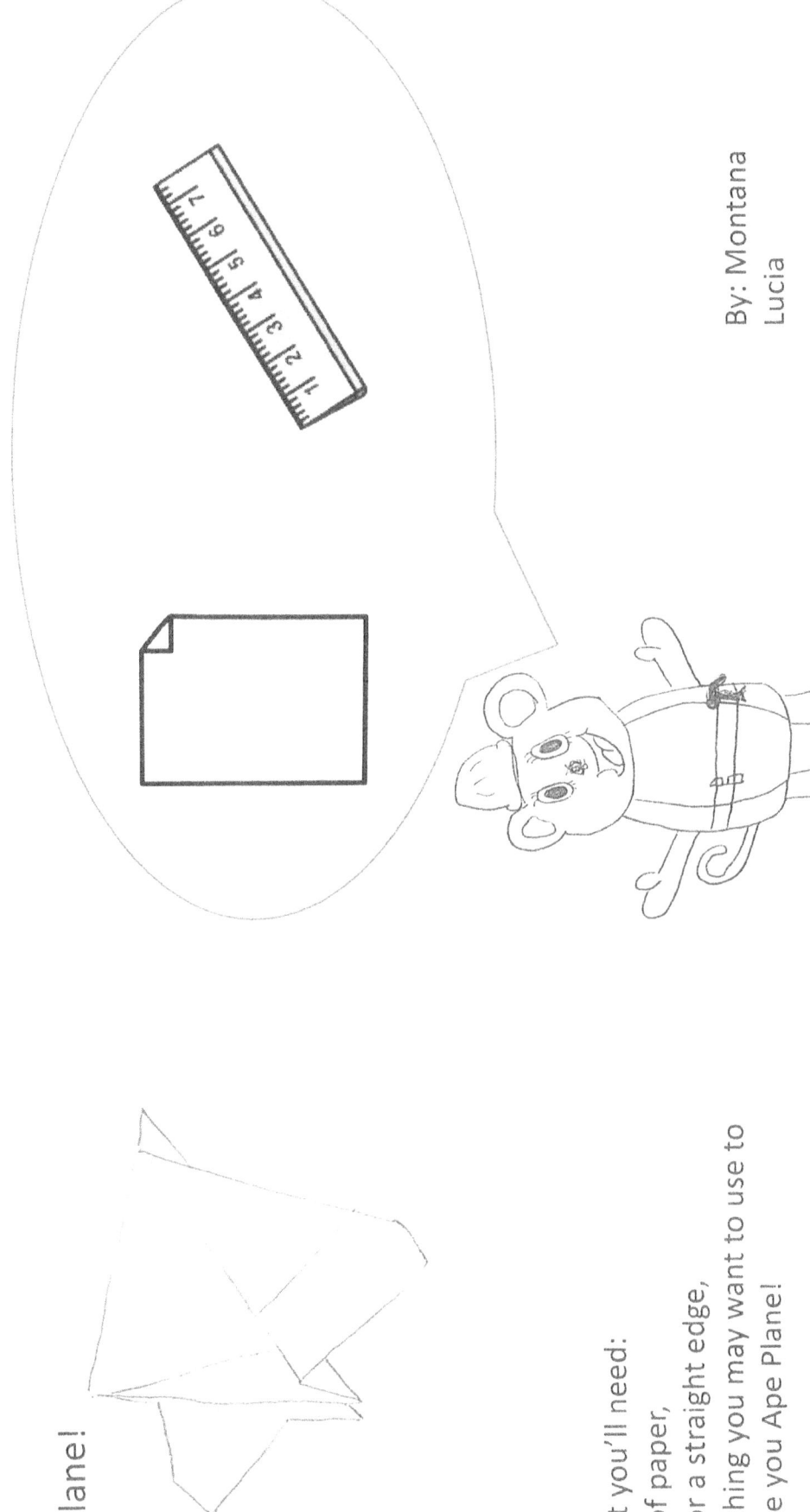

The Ape Plane!

By: Montana Lucia

Here is what you'll need:

- a sheet of paper,
- a ruler for a straight edge,
- also anything you may want to use to customize you Ape Plane!

Step 3: Fold along the line both sides down on opposite sides.

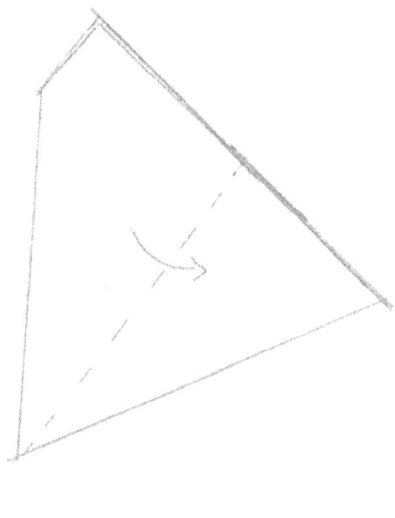

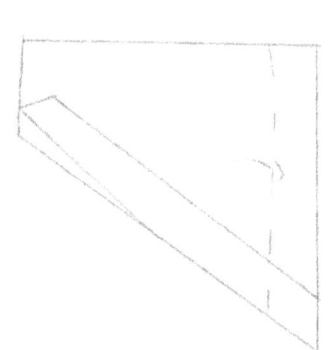

Step 4: Fold the previous pieces down on the line

Step 1: Fold corner to corner, some overlapping will occur.

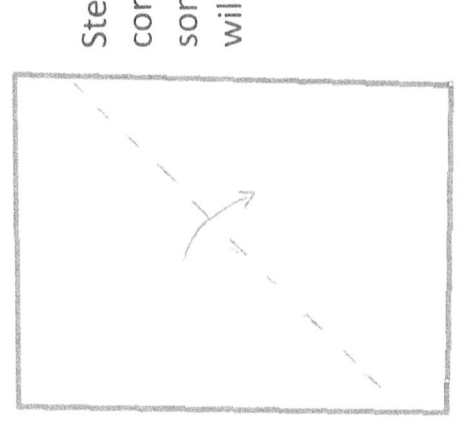

Step 2: Fold on line 'A', then line 'B'.

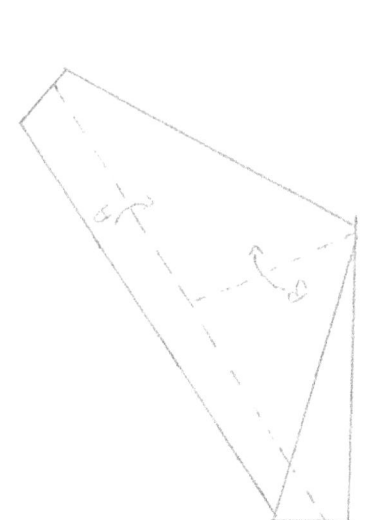

The Squirrel

By Mikaela Raymond

Step 2:
Fold one corner down to the center line.

Step 3:
Fold the other corner down on the opposite side.

Step 5:
Fold the pointed side to make a square.

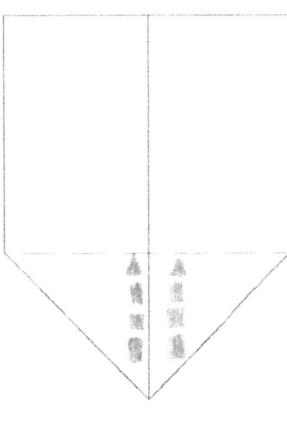

Step 7:
Repeat the fold on the other side.

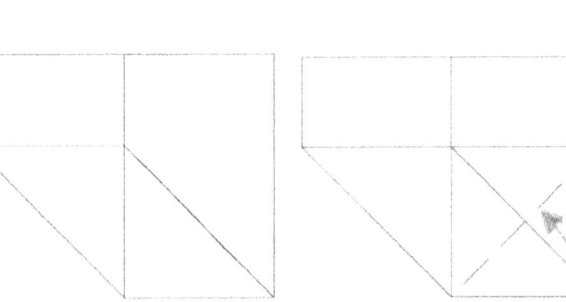

Step 4:
Unfold along the center line.

Step 6:
Fold along the dotted line so the corner meets the center line.

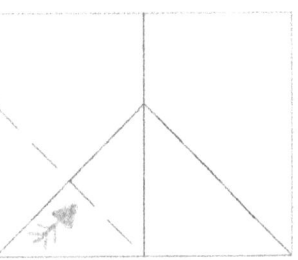

Step 9:
Fold along the center line so you can see the different folds.

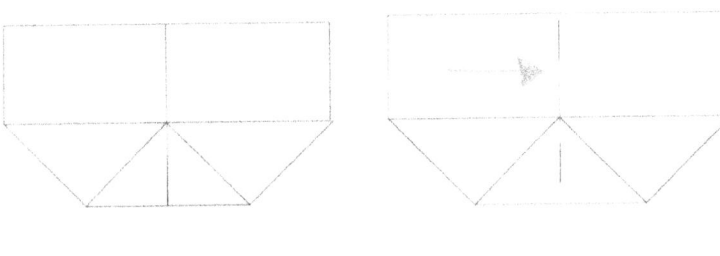

Step 8:
Fold the pointed side down to meet where the other folds cross.

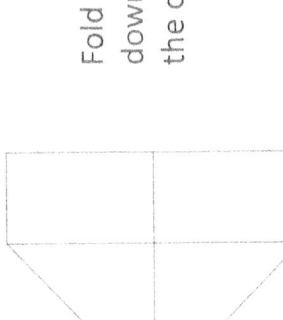
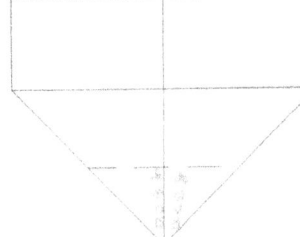

Step 11:
Fold the other side out to match the first wing.

Step 10:
Fold one side on the dotted line to make a wing shape.

Step 12:
Unfold the wings to finish the plane.

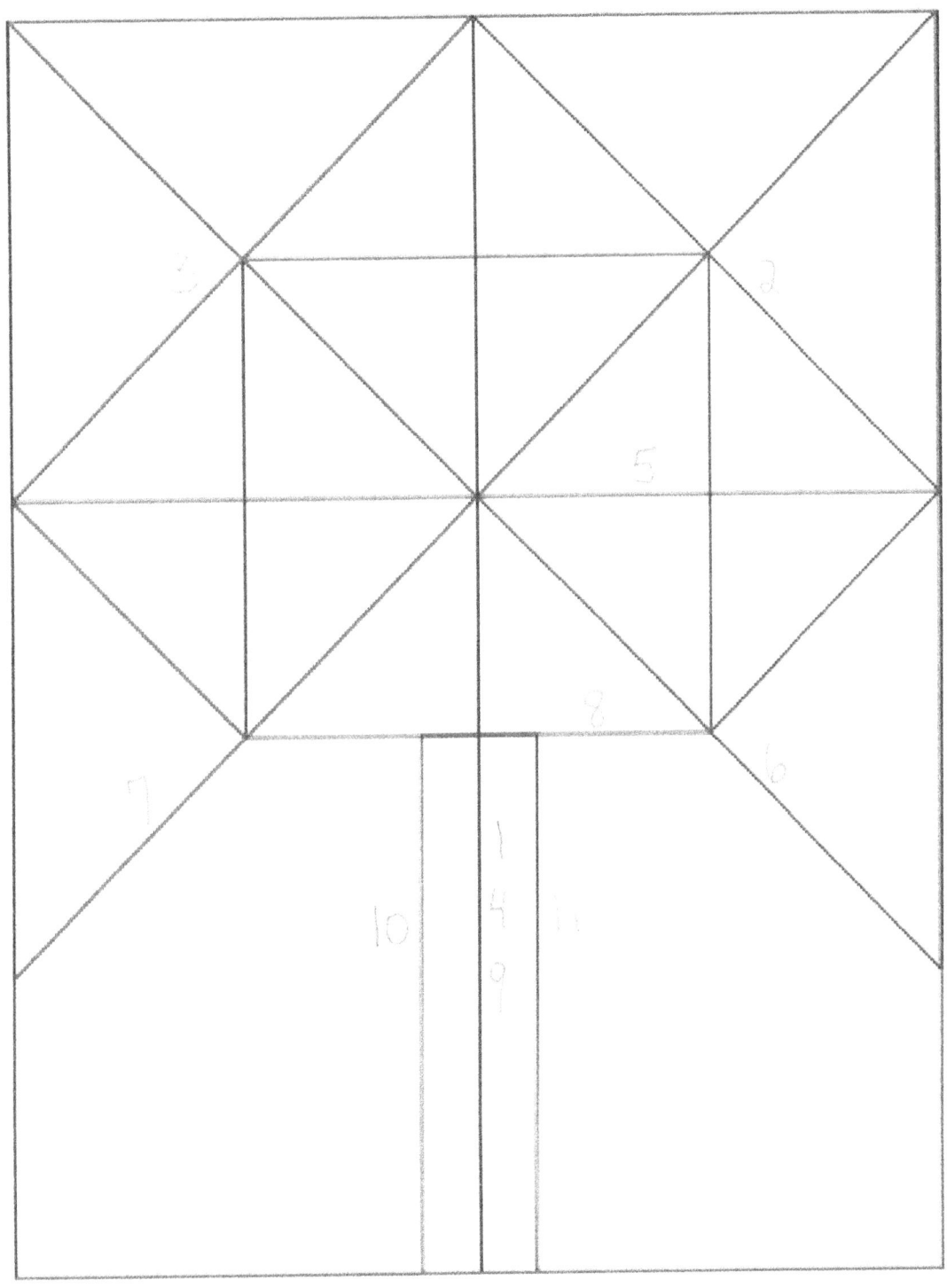

THE SQUIRREL

Space Shuttle

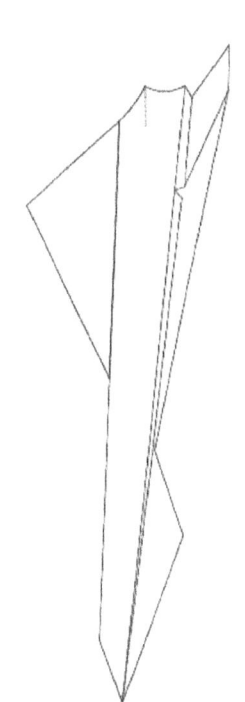

Paper aircraft designed by
Michael Weinstein

Illustrated by
Logan Hebert

-Legends & Symbols-

- Fold line/Crease line

- Distance

- Fold direction

- Fold back & forth (or side to side)

- Fold behind

- Fold then unfold

- Unfold

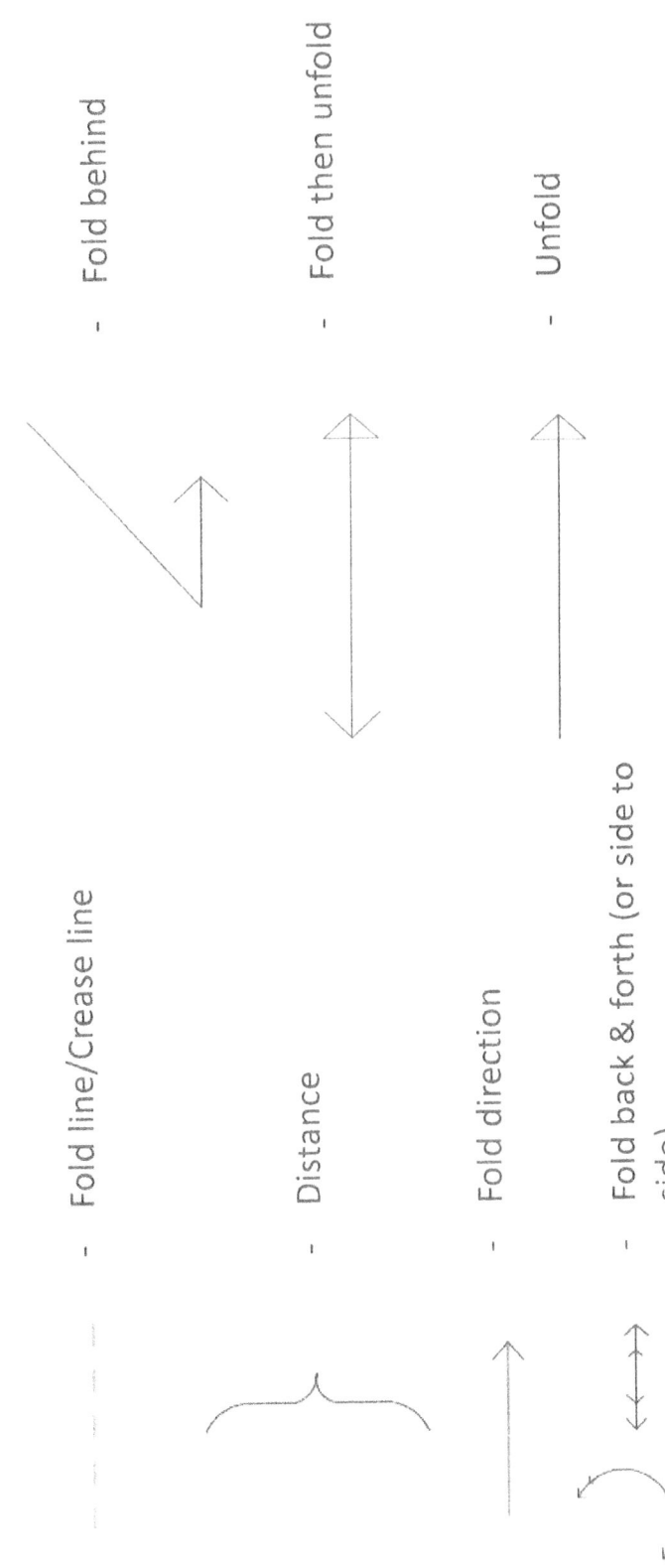

Step 1

Fold the paper in half, then unfold.

Step 2

Fold diagonally at the dotted lines, then unfold.

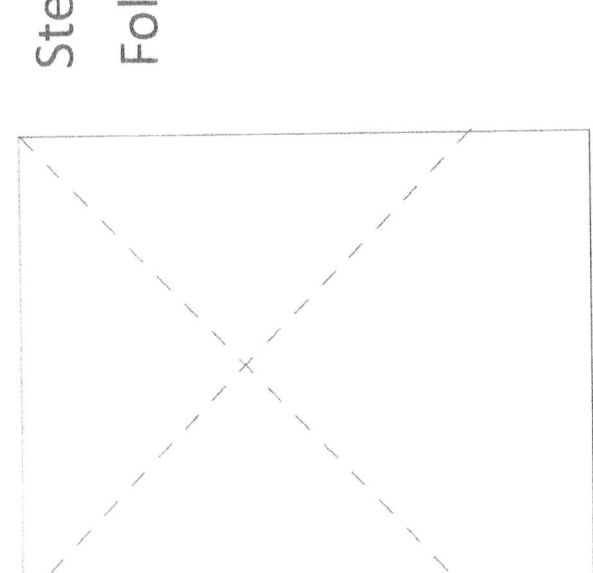

The resulting folds should look like this to add the crease lines.

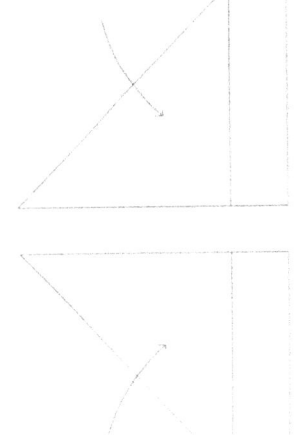

Step 3

Now, using both hands and fold the paper up and in toward the middle.

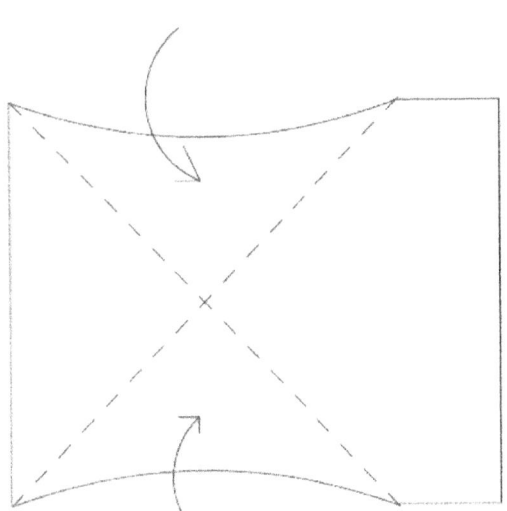

Step 4

This is how the paper should look from the given fold, Now press down the sides and middle at the uncreased bends and.....

Step 5

The resulting fold should look like this.

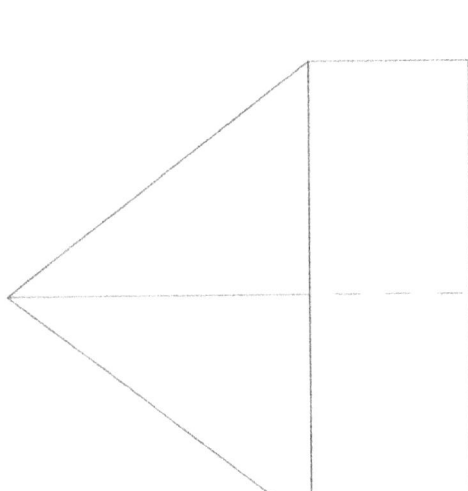

Step 6

Slide your right hand under the right flap, as shown, and flip the right flap to meet with the left, as shown. They should line up evenly

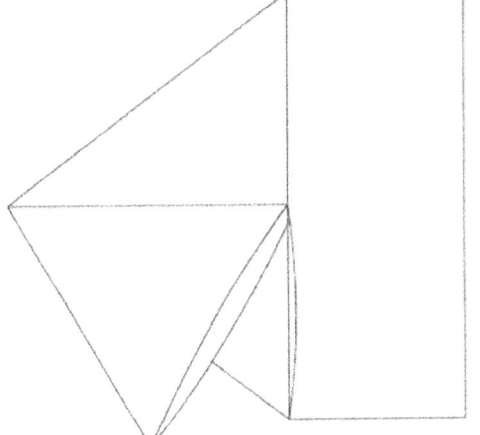

Step 7

These should be the results. Now, grab the right flap, which will be called flap X, fold down the center on the dotted line to meet up evenly with point A and after....

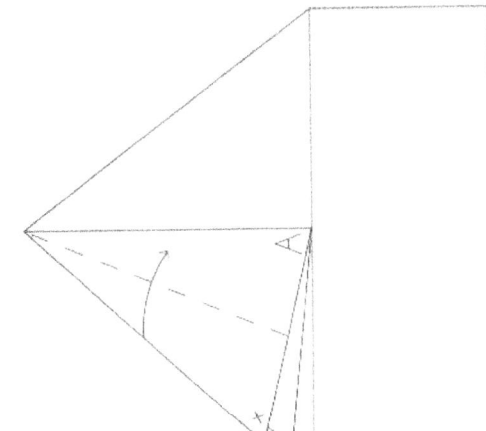

Step 8

Flip flap X over back to its point of origin. This should be the result. Now, notice edge B, and we are no longer calling A 'point A' but now it is Flap A. Now repeat steps 6-8 with Flap A.

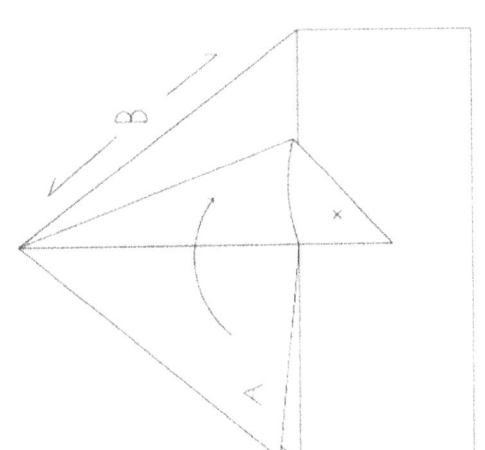

Step 9

This should the result. Next is to flip flap X to....

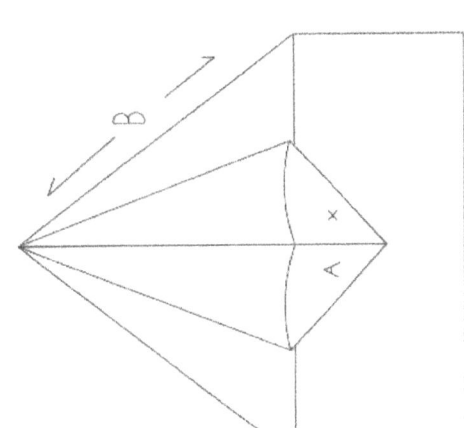

Step 10

Flap A. This will give you room for Side B. Like done with Flap X, fold Side B in half down the dotted line so that Point C meets with Point A (to a degree it will meet).

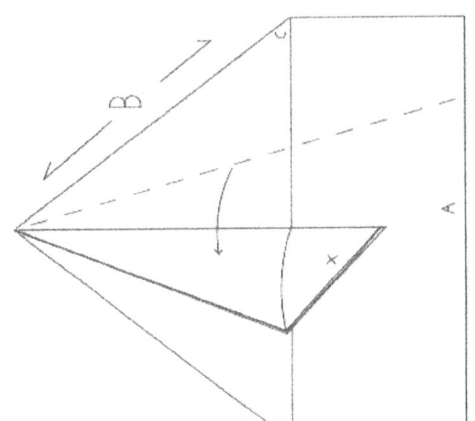

Step 11

This is the new Side B, it should look like this. Do the same to the left side.

Step 12

Now that both sides are done. Spread apart Flaps A & X like so

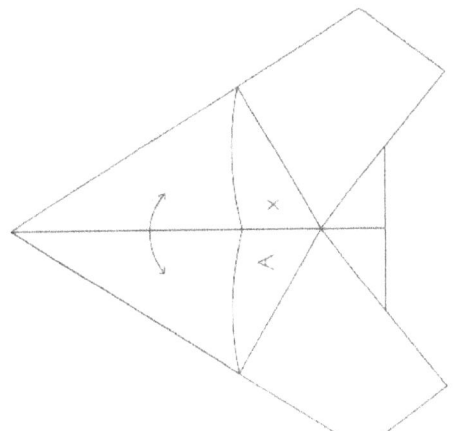

Step 13

Move Flap X back over to the left side and do so as illustrated. Side B should meet perfectly with Side Z. Repeat to left side.

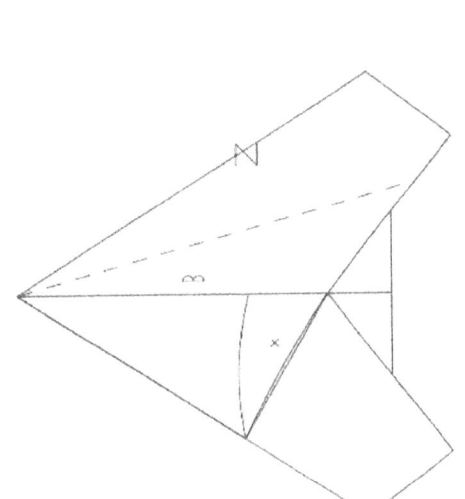

Step 14

Fold the right wing along the dotted line and precisely to the measurements given.

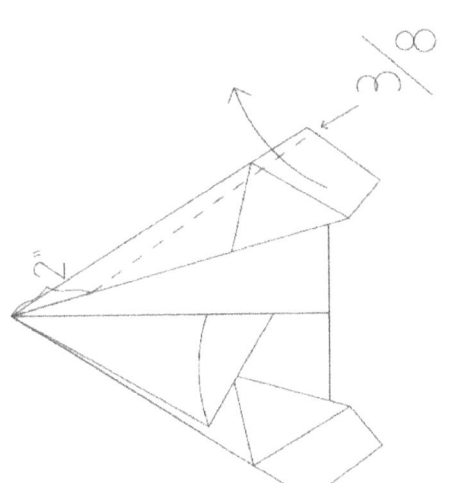

Step 15

This should be the resulting fold, now repeat to left side again.

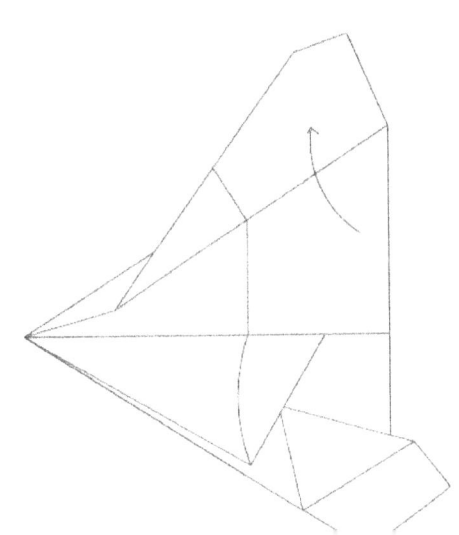

Step 16

Now, fully open Flap X to the left and fold along the dotted line so that Corner X meets with Corner H.

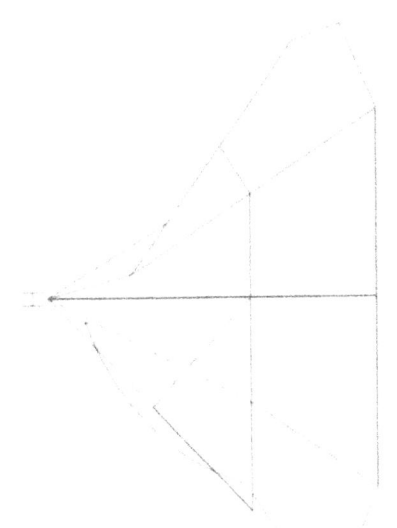

Step 17

This should be the resulting fold. Repeat on left side.

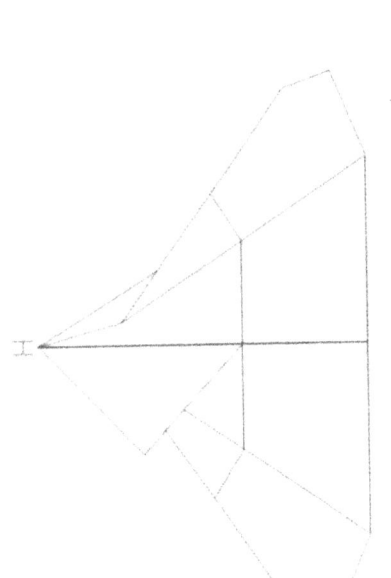

Step 18

Above the dotted line is the nose, fold that down towards you along the dotted line.

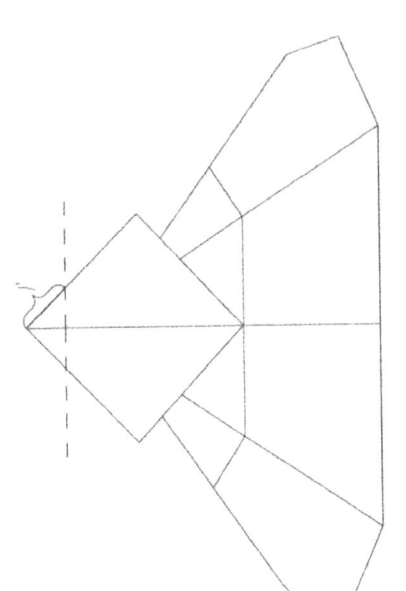

Step 19

This is how the fold should look. Now, fold both sides of the aircraft away from you.

Step 20

The plane should now look like this. Lift the nose up and fold along the dotted line pinching the belly and folding the wings downward. Fold from side to side a few times.

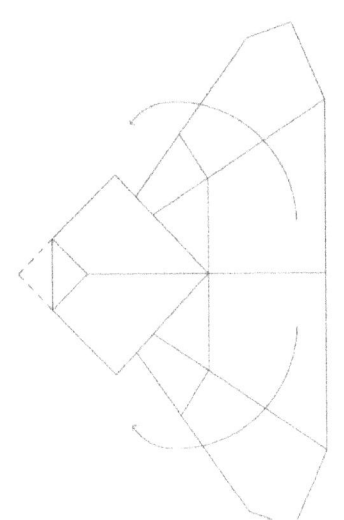

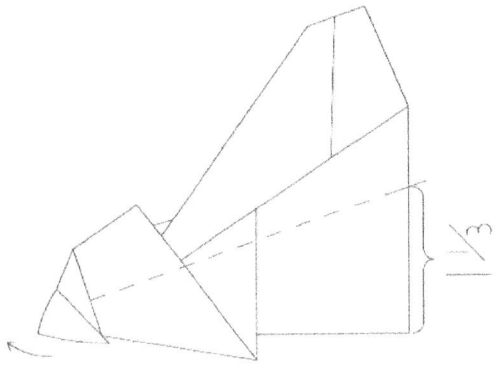

1/3

Step 21

This should be the result . Now fold the wings up and below the fuselage, which is the 1" marked area, and sink the belly upwards to make the rudder for the aircraft. The belly is not illustrated but is found under the wings, the same belly you pinched to fold the aircraft wings and fuselage down in the last step. Leave some of the belly under the aircraft too. Fold the wings up and down to keep it centered.

Step 22

Finish this off by spreading the fuselage and doing a light fold on the dotted lines to make the engine, do the light fold by protruding the area outwards and pinching lightly until the area sticks out. And you're done!

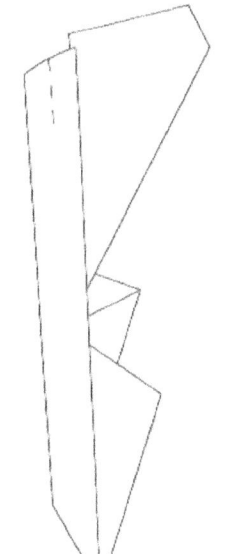

www.ingramcontent.com/pod-product-compliance
Lightning Source LLC
Chambersburg PA
CBHW080832170526
45158CB00009B/2551